Roshan Cipriani

RISE

BE BOLD~ BE TRUE TO YOURSELF~INSPIRE OTHERS TO LIVE

INTRODUCTION

I was obsessed with the idea of being a writer. I wanted to make a difference. I wanted to travel and speak and write books!

I didn't know how I'd make it happen. It was so far from what I was living. And I was scared!

That was the exciting vision for my life. I knew that in order to succeed one day my belief would have to become bigger than my fears.

I didn't know what the journey would look like or where it would take me, but one day in the depths of my despair I made the decision to RISE.

On this path, I've had to embrace uncertainty and some of my biggest fears. I've fallen and gotten back up again and again.

That moment when the Universe gifted me with the ability to decide to RISE and do what I believe is my real purpose, has been the greatest gift my Creator could have given me.

"Phoenix rising from the ashes."

This phrase popped into my head recently, and I wasn't sure why until I did the research.

As the story goes, the phoenix is a mythical bird with fiery plumage that lives up to 100 years. Near the end of its life, it settles in to its nest of twigs which then burns ferociously, reducing bird and nest to ashes. And from those ashes, a fledgling phoenix rises – renewed and reborn.

And now I get it. This is the story of my life in the past few months – especially the part about burning ferociously.

Life presented me with some challenging circumstances that left me just hanging on. My experience is not unique – it happens to all of us at some point, it's a part of the human journey.

However, this was my time, and I'd like to share with you what I've learned and am still learning along the way.

When I look back I see that I had very little control over what happened. I was so busy trying to make things happen that I overlooked what was actually happening. I have been going through something I didn't expect and which I had no control over.

Some days are better than others and although I still can't see the light at the end of the tunnel, I do know it is there. Anything that brings us to realize presence is a benefit.

I am just now coming to the realization that I have to go with the flow and that the anxiety is trying to tell me not to resist. I'm grateful for the reminder to 'go with the flow' and allow things to take their course.

Power comes in being awake to these cycles, kind of like the seasons, we need to know they are flowing past, if we swim against this flow, we will wear ourselves out and drown.

I truly believe that each of our experiences becomes more meaningful when we are able to share them with others. Suddenly your struggle is transformed into a greater purpose if it can be used to inspire, give hope to, or relieve the pain of others.

Yes, things go up in smoke sometimes. They – we – burn up. It's a holy fire that has the potential of giving birth to the new.

Swimming against the flow makes things harder, for sure. What I found in my recent experience is that part of the process was to resist. I too have experienced the same feeling of complete and utter hopelessness, and yet still trying to fix things on my own.

When I woke up to the fact that I was resisting, the mind takes over, trying to understand, interpret, and control. Going through a tough time is not a mistake, it is just another part of the process, a chapter in life, so that even the resistance isn't wrong. It is true that once we finally accept what is happening and let the anger and resistance "fall away" if you know what I mean…that is when the healing and the "rising from the ashes" can begin. There is comfort in what is familiar, even if it no longer serves or reflects our personal truths.

 I found it easier to let go when there is something better, that being a meditation or concentration regiment. Meditation is a wonderful supportive practice to this realization. Meditation and the practice of chanting and writing Mantras have proven to be the best thing in my situation.

The threads that run through these experiences for me are self-acceptance and surrender. I'm very solution oriented myself, but you quickly learn that you can't push the ocean, it flows by itself in its own timing.

I tried many solutions. All of them went nowhere because, as I eventually discovered, what was needed was self-acceptance and surrender.

You cannot control the uncontrollable.

Sometimes all we have left, at least it seems that way at times, is the trust or willingness to keep going no matter what, because when you're in the thick of things, you can't see a way out.

Sometimes, we have got to go through the fire. I have had a few experiences in my life already where I finally realized that there will be times in life where there is no way around the fire.

The greatest healing comes through complete acceptance of everything as it is – even those experiences that don't fit our belief systems or those that make us want to run away. Everything that comes our way holds the potential for transformation – especially the fire.

I found that I held to these thoughts, and they are my personal truths - "I do not have to have the answers and be able to work it all out, right now, while the craziness is unfolding. I will not pretend to know what I don't know."

Often I felt like one of those birthday candles that will not blow out, continuously igniting at the seemingly wrong moments.

I have learned that all I can really do is "Relax."

Everything will take care of itself in its own time. We love to fix and know and have control, but sometimes we just can't. Relaxing eases the stress and tension.

I am happy again. I have chosen happiness over fear in my heart.

I'm letting go of guilt and depression which has tortured me and I am learning to forgive and love myself again. It has been a long road and it is ongoing.

I believe everything happens for a reason, and I try feeling blessed for all that is there, rather than what is not.

I have found that worst time of the day is when I get up in the morning and when I go to sleep at night. There are profound moments when you feel the emptiness and the loss…it is a terrible feeling but it does pass.

As crazy as it might sound, the key to rising is to move beyond your story. Introspection of negative thoughts is also a form of meditation. So you have to learn positive thought processing.

To start, you have to begin to learn to lose interest in your old thoughts. They have been coming at you for a long time, and you have a habit of identifying yourself with them. Stop it.

Remember in pain or in fear you are not your circumstances and you are not limited within a horrible story.

Just know it's actually ok to feel bad and not have it all figured out and to not know when you may actually figure it out.

To trust in the process is all there is.

This is the opportunity to be present, here and now – to do the next thing that is right in front of you, and to take very good care of yourself as best as you can.

Daily thank God for even a small ray of hope that at some point, all is resolved and settled and you can feel peace and happiness without the flames. Put out into the world what you want, and that will be your experience.

Acceptance does not require me to like or agree, but rather is simply my acknowledgement of what is.

My body is a far more reliable barometer than my mind for giving an accurate real time reading of my acceptance of the current moment.

By listening to my body, I'm often able to recognize discomfort is letting me know that it is actually fear or worry of the future that is hindering my acceptance of the now.

 Even though what is actually happening now is not fear filled, I have been so used to the feeling of dread that I retreated there without any reason to. Wellness is an active process of becoming aware of and making choices toward a more peaceful existence.

So that the interaction of body, mind, and spirit with appreciation that everything we do, think, feel, and believe comes from a place outside of our experience.

Recently I have learned the true meaning of acceptance. I never in all my life thought I was going to experience certain emotions and be dealing with confusing situations, as I'm doing right now.

Nothing is permanent, everything changes; and of course as you live and observe other peoples drama, you think you'll never experience what they're living, whether it be a positive or negative situation.

You think, "That will never happen to me."

When something happens that's out of your control, you quickly learn that no one is exempt. Often you think that if you gave everything that was in your power, and that if you persisted enough only good should have been in your path. We have a belief that things could and should be different than they are. This creates irritation and disappointment.

Eventually you begin to realize that change begins with acceptance of reality. This leads you to accept your powerlessness and your limitations. Things are happening exactly as they are supposed to. There are no mistakes in the Universe.

When we fail to acknowledge the truth about what is, we get stuck in arguments and actions that cannot succeed because success cannot come from a misunderstanding of reality.

Sometimes, that's all we can do. There may be nothing on the outside that we can change, but acceptance of a situation brings peace of mind and allows us to enjoy the moment. You have this right.

You may not believe you have any rights if yours weren't respected growing up. For example, you have a right to privacy, to say "No," to be addressed with courtesy and respect, to change your mind or cancel commitments, to ask people you hire to work the way you want, to ask for help, to be left alone, to conserve your energy, and to not answer a question, the phone, or an email.

These are often the only types of things that you are able to make decisions on when you are in the middle of a crisis. When you understand that it is your right to be in the moment that is when things begin to change. Things will ALWAYS change…don't hold up the progress by your resistance.

The greatest remedy in the world is change; and change implies the passing from the old to the new.

When change feels like walking off a cliff blindfolded, people will reject it, often preferring to remain mired in misery than to head toward an unknown.

By definition, change is a departure from the past. Old wounds re-open, historic resentments are remembered — sometimes going back many generations. Try to live moment by moment.

Wherever possible keep things familiar. Remain focused on the important things like breathing and sleeping. The universe is already orchestrating everything else.

You have mainly two tasks.

You are supposed to starting thinking for yourself. And the other is to break out of the limited expectations set for you by others.

You are meant to start discovering who you really are and what you are meant to do with your life. And the people and things around you can really help you in that, or hurt your efforts.

I want to show in this book that it is never too late. This experience can change you for the better. You can become bold, true to yourself and inspire others to live.

Some people leave it up to their deathbed, but even then you can discover who you are -- although it makes it a bit tough to reveal your answer. But even then, you have heard stories of people who have become a higher, gentler, more loving version of themselves -- have found an oasis of peace that others envy -- in their last days. Most people feel a bit awkward at first with change, simply because the universe is doing all the work and rearranging your life the way it is supposed to be, instead of how your brain or culture or family may expect it should look.

When change is out of our hands, we become frightened -- no scared stiff.

However when you have built a life on a bad foundation, on a false idea of who you are and what is possible – you just need a big push to tear it down and build a new one. What is so unique and wonderful about discovering yourself is that help is provided by the universe. It does all the work you just have to move out of the way.

The basic outline for our life was planned (pre-destined) before we were born to give us the right opportunities for growth. But we don't have to follow the pre-destined route if we don't want to. A life plan can be compared to a maze, and our life's purpose is to complete the maze.

We don't know where we are going or where we will end up, so all we can do is make our way through life trusting in our Creator. If you chose not to follow the predesigned route, it changes nothing, as no matter what road you take you will still end up at the preordained destination. You will only inconvenience yourself. The universe and or our guardian angels can only guide our destiny to the extent that we allow them. The sensible option is to give them complete control because they are far more advanced than we are .Yet most people are egotistical and think they know better.

Karma is a Sanskrit word meaning "action", but is generally understood to mean the consequences of one's actions. The word "karma" is commonly used to indicate bad karma, and the word "merit" is often used to indicate good karma. The law of karma is best described as "cause and effect" because every action (or cause) has a corresponding consequence (or effect).

If you plant good causes you will reap good effects, and if you plant bad causes you will reap bad effects.

The Bible describes karma in Ezekiel 18:20: "The righteousness of the righteous will be credited to them, and the wickedness of the wicked will be charged against them".

Many people believe erroneously that wealth and abundance are gifts from God as signs that he is pleased with us and has blessed us.

This belief is completely wrong; God does not have any favorite beings and our financial status is of no concern to him.

Wealth is the result of good karma or merit; the result of "good investments" we get due to our ancestors behavior.

This is why we never know how or when the karma we create today will manifest its consequences. This is because karma works behind the scenes – we only see the effects in the physical world and have no idea of the causes that are operating out of sight in the other worlds. Karma and destiny are woven into our lives to the extent that the average person barely notices their effects.

I the Lord thy God am a jealous God, visiting the iniquity of the fathers upon the children unto the third and fourth generation.

Exodus 20:5 , Deuteronomy 5:9

Their blood shall therefore return upon the head of Joab, and upon the head of his seed

forever.

1 Kings 2:33

Visiting the iniquity of the fathers upon the children unto the third and fourth generation. Numbers 14:18

Wherefore hath the Lord pronounced all this great evil against us? ... Because your fathers have forsaken me, saith the Lord.

Jeremiah 16:10-11

Thou ... recompensest the iniquity of the fathers into the bosom of their children after them.

Jeremiah 32:18

Now we see that within the stories of even the book of Genesis in the Bible that which happens to the earliest characters in the narrative will be repeated by their children.

This is why we have to pay extra special attention to what's going on at this early phase of the Bible, because here is where the patterns are set. In this early narrative is the map, and the guidebook for the future.

Karma is self-balancing divine justice; it is 100% fair and absolutely infallible. It doesn't matter if a criminal seems to "get away with it", because there is no getting away with anything. It is impossible for us to understand the intricacies of karma because it is controlled by intelligences that are far superior to us.

The best way to deal with these events is to accept our fate and let it work itself out. We should accept the hand that life deals us and go with it, because if we resist we will only make ourselves more miserable in the long run. We must however do everything we can to alleviate all suffering even our own, because every good deed contributes to a better world for everyone. Relieving another person's suffering not only helps them; it generates good karma or good righteousness to be credited for us and sends out good vibrations and frequencies into the universe. Does any of this take the Most High God by surprise?

Absolutely not.

 First, since God is sovereign, that is, completely in control of everything, nothing happens that God is not ultimately the cause of. Second, since God knows everything that is going to happen, nothing can ever happen that --

1) God doesn't already know about.

2) God hasn't personally caused to occur.

<u>Get Knowledge</u>

Knowledge causes your brain to change, making you change physically. This is all by design.

As you read this book, the physical make-up of your brain is actually changing, and therefore your environment is changing. You are becoming a little different physically. Anything you learn changes your brain a little bit. The more time you spend with something; the more impact something has on you. Your brain and your environment actually changes physically as you read this book.

It is also no coincidence that you are reading this book right now. Everything even this is already outside of your planning and control. You can only do one thing that matters, and that is to see the truth and live the life that is being given to you.

You are only truly free when you know for certain: You are not in control.

Introduction

Index page

<u>Chapter One.</u>

Relax And Let Go

<u>Chapter Two</u>

The Gratitude List

<u>Chapter Three</u>

Living One Day At A Time

Chapter One

Relax And Let Go

"Be still, and know that I am God"

(Psalm 46:10).

The word translated "be still" comes from the Hebrew term *raphah*. It refers to that which is slack, or to let drop, so we need to drop our hands, go limp, relax, and "chill out."

So as your world crumbles around you, the call from the Biblical word is: don't flinch in faith in God, just stand still.

The number one reason you're under stress is because you're in conflict with God. You're trying to control things that only God can control. Not only are you going to lose that conflict, but you're also going to be tired. You need to be humble.

That is a Universal Law.

You see, in my case my life crumbled a year and a half ago. I remember being on my knees one day, tears streaming down my face. I humbled myself before God. I was so tired and fragile. I feared that I no longer had the strength to push on. I can remember crying and telling God that I was sorry for all my mistakes. I prayed to God that day and said "God, I am so tired. I do not think that I can do this anymore. Please God, tell me what you want me to do? Help me God, please!"

In Habakkuk 1:2, Habbakuk asked the following of the Lord:

"How long, O LORD, will I call for help, and You will not hear?"

This was one of those big, deep, tough, "life" questions that Habakkuk was asking God.

In the midst of all the chaos surrounding his life, he was basically asking, "God, where were you when I needed you the most?"

I fully understood this man.

I was tired of suffering and praying and having God appear to be silent.

God said to Habakkuk, "Even though you don't think I'm listening, Habakkuk, I am working on a plan that is so much larger than you." The truth is, even though Habakkuk couldn't see it, God was working the whole time!

So the law of cause and effect is not punishment, but is wholly for the sake of education or learning orchestrated by our Creator.

A person may not escape the consequences of prior actions, but will suffer if the conditions already ordained are ripe for suffering.

Ignorance of the law is no excuse whether the laws are man-made or universal.

THE LAW OF HUMILITY

What you refuse to accept will continue for you.

Every feeling, every thought, has been prepared especially for you even the ones you think are evil.

Psalm 139:4

For there is not a word in my tongue, but, lo, O Lord, thou knowest it altogether.

Psalm 139:16

Thine eyes did see my substance, yet being unperfect; and in thy book all my members were written, which in continuance were fashioned, when as yet there was none of them.

God created evil for a purpose. So naturally He will not make evil void or there would have been no purpose in creating it in the first place.

Isaiah 45:7

I form the light, and create darkness; I make peace [good], and CREATE EVIL; I the LORD DO ALL THESE THINGS.

Out of the mouth of the most High proceedeth not EVIL AND GOOD?"

Lamentations *3:38*

"...I [GOD] will bring EVIL from the north, and a GREAT DESTRUCTION"

Jeremiah 4:6

"...Hear, O earth: behold, I [GOD] will bring EVIL upon this people...

Jeremiah 6:19

"And he said, I [the lying spirit] will be a lying spirit in the mouth of all his prophets... and He [GOD] said, ...GO forth, and DO SO"

I Kings 22:22

"He [GOD] turned their heart to HATE HIS PEOPLE..."

Psalm 105:25

"...Thus said the Lord; Behold, I FRAME EVIL AGAINST YOU, and devise a device against you..."

Jeremiah 18:11

And it came to pass, when the evil spirit from God was upon Saul, that David took a harp, and played with his hand: so Saul was refreshed, and was well, and the evil spirit departed from him.

1 Samuel 16:23

If we accept the principle of cause and effect in Nature, and of action and reaction in physics, how can we not believe that this natural law extends also to human beings?

A finite cause cannot have an infinite effect.

Eventually your suffering will cease.

Job 2:9 states "Shouldn't we accept the bad as well as the good from God?"

Job realizes the nature of God and creation, and teaches radical acceptance: facing the truth of what is as the first step toward positively engaging with what is.

Reality acts according to its own laws; it doesn't try to meet our expectations. While delusion is often characterized by complexity, the truth is simple.

The greater our expectations and the stronger their disconnection with reality will be the deeper our suffering and resentment.

If you're still shouting, and saying things like – "I refuse to accept this is the way things are; I don't want to and I won't put up with it, even though there's nothing I can do about it", you are acting like a child having a tantrum.

You could start shouting, "Why is this happening to me? How have I ended up in this mess? I don't deserve this"   You could fight against the daily trials and curse your life. Such actions are non-acceptance: you're refusing to accept the situation you're in.

For some reason, you've ended up in this particular situation. Why this exact destiny has landed on your doorstep isn't important. There's no point thinking about it, crying or shouting at the seeming injustice of it.

You're in it and there's nowhere to escape to.

 It therefore makes more sense, from the point of view of survival, to see the reality of your circumstances.

Acceptance doesn't exclude thoughts and making plans intended to arrive at change in a situation. It doesn't mean making no effort to influence your circumstances.

Acceptance signifies wordless agreement with the way things are NOW, coupled with a willingness to meet whatever might surface and to act appropriately when necessary.

People

People may blame you for something you didn't do. These are the realities of the human condition which must be expected in order to free you from feeling pointlessly irritated by the words of others. Usually people who have never been in your circumstances feel the greatest need to offer banal explanations and advice.

At first I was annoyed and then I became angry, and then I was hurt. I found it very hard to come to terms with these people and their unwanted advice. However I soon realized that there was no point thinking about what they said if it was right or wrong.

 What did it matter?

What happened had already happened, and was still happening and nothing would change that it had already occurred.

Yes, people behave this way; was this really a new revelation to me?

No, it was not.

People behave emotionally; they can be opinionated and rude. You cannot control this.

It's the way people are.

Acceptance allows you to remain calm, to free yourself of many unnecessary and unproductive thoughts and feelings.

Get rid of your expectations that people should treat you in a particular way as it is unrealistic, and stops you from concentrating your energy on you.

Get rid of your ideas that you should be better than everyone else, that you have to satisfy others' expectations, or that you need to prove your innocence and value to people.

<u>Know one thing as a fact: you're NOT always going to be the way you are right now.</u>

Rid yourself of the expectation that you can control anything including others opinion of yourself.

Our tendency is to try to compensate for our fear of being thought to be a victim or inferior by trying to control how we are seen by people and this manipulation can also be detrimental to our own minds.

The reality you want will be given to you at the most appropriate time, according to the universe – not on your timetable, or due to the recommendations of other people.

The universe has laws that require a situation to literally turn around again.

Change your attitude against your circumstances and realize even though you can't see how, everything it is still working in your favor ultimately.

<u>Don't abandon what you want. That is the key.</u>

 It might look like your future has been destroyed but don't mistrust or lose faith in Universal Laws.

Do my words seem crazy or impossible in the situation you are in right now? Yes, reality might be a harsh place.

Nothing in this life is permanent. I used to strongly dislike hearing this statement, particularly since I greatly feared any change.

However, now I find it's actually quite reassuring. Nothing; not even my old life's destruction, and this crisis will be permanent.

Understanding leads to acceptance. Acceptance always leads to change.

"When you change the way you look at things, the things you look at begin to change."

(Wayne Dyer)

Many people are unaware that the past, present and future are in a constant state of change. You may wonder how come you are unaware of let's say your future and your past changing.

Well, when reality shifts, so does your memory of many events. As far as you are believe and think, yesterday was the same as today, because the remembered yesterday matches exactly with your remembered today.

This is why when two people try to recall what happened on some previous event each person usually tells dissimilar accounts of what they experienced together.

In many of these cases, both people experienced the exact same thing, yet each person recalls it differently.

As a person's individual reality shifts, one person experienced the event one way while the other experienced it in a totally different way. You see while these two people did share the same event because of the physical reality changing nature each person, now, has moved into a different version of the event and to a different recollection.

Reality Shifting Dimensions

<u>The future the past and the present change constantly in a non-physical environment</u>.

It is only in the physical that things become stagnant for a time. When a change occurs in the physical environment our brains can perceive it almost immediately.

I was shopping and running errands. Two different people at the pharmacy greeted me, and said how odd it was to see me again after just having just rung up my purchases a few minutes ago. I had not been to the store before, and definitely not in a UPS uniform that I was apparently seen in.

Neither had I had conversations with other customers as these two people indicated had happened.

Remember the scene in the matrix where the same cat walks by twice? I saw the same thing happen.

I was at the sink washing dishes and I saw my friend walking down the back stairs, wearing a light blue shirt. In another second, he was right there standing next to me dressed in gray.

I was startled and asked, "Hey, how did you get back here so quickly?" "Weren't you wearing a blue shirt?

He gave me a puzzled look.

Then moments later I saw him go out the back door where he walked down the same back stairs. He was wearing a light blue shirt.

It was like a movie scene being re-played, some sort of time overlap.

These types of events are not unusual and are being reported as having happened to many different people in many countries.

How could this be happening?

Physical reality was created to be stable. This how we are able to observe, and learn from the familiar things around us. When we start to notice that something that was familiar to us is no longer as we remember it to be just yesterday, it is a clue that your reality has changed.

It's not demonic. It's not that you are forgetful.

It's quantum physics.

In fact, if your reality change becomes great enough, it can take days for your new memory to seem real to you. You are likely to remember things that happened in an old reality but do not occur in your new reality.

This is because the brain is not static, rigid and fixed. Brain cells are constantly and continually renewed, remolded and reorganized by our thoughts and experiences.

However after a while most people will just excuse it away as some mental lapse or being mistaken, because everyone else seems to do just that. We think maybe I'm getting old and becoming forgetful. However, there is another explanation.

The events described in this book underscored moments when subjective reality overlaid objective reality to determine my new experiences. When that happened, the future easily surfaced completely changed from what I knew in the past. These events occurred automatically, without any effort on my part, and regardless of logic or prior beliefs.

What we call time -past, present, future – stopped being sequential and formed its own pattern with things I could not explain.

Right now, as you continue to concentrate and read this page, you might have forgotten about the crisis and situation you are in.

So in your reality they actually ceased to exist at that time.

For just those few moments that you were not thinking, and unaware of many things in your life, great changes have already occurred.

Our attention brings everything to life and makes real in our reality what was essentially unnoticed or unreal before.

The biggest reality shift that I have ever experienced to date happened when I began my prayers and meditations by focusing not on avoiding violent death and problems in life, but prayers and thoughts to move my life to a better reality or dimension. I share this with those people dearest to me.

This also has the added benefit that whenever we experienced changed reality from the present familiar reality, we have seen the same changes in the physical environment and can point out and discuss the changes.

When you do it on your own, it can be easy to think you are just imagining things when you see objects and buildings appearing and disappearing within a day. Often you will see familiar people in unfamiliar roles or environments. This is another sign.

Is it possible that one person's thoughts can change physical reality?

So why don't more people notice when these physical changes happen?

From my own experiences and from those of so many others who are experiencing reality changes, I believe there is an explanation for how it can happen. In the idea known as the Multi-Dimensional Universe Theory in quantum mechanics, modern scientists have a general idea about reality. Each reality field is a separate dimension with its own timeline experiences and outcomes.

It comes complete with its own Universe, dimensions, space-time continuums, frequency domains, and created entities. All these dimensions exist inside each other like the rings of an onion which is the "Quantum Matrix".

They believe that you exist simultaneously in a certain number of these rings or dimensions. If you lead a life that is very sheltered, with little contact outside of your town, you might exist on a few hundred thousand timelines or less.

If you have a life of interaction with people and travel often always making new experiences and meeting new people, you might exist on a million or so timelines.

When you do anything, your conscious self reaches across to another dimension. Your thinking transports you to a new place.

"You" are still in the first reality timeline or dimension. However you are only minimally aware of the "you" on the new timeline and dimension.

Often in living your life events will give you a reason to question things and events.

My daughter and I had a harrowing experience that left us profoundly shaken and disturbed. I searched for answers to why I felt uneasy and in unfamiliar circumstances, with everything I thought I knew now so different than I remembered it.

I read up on muscle testing for determining hidden universal truth.

In using these techniques, to determine those subconscious truths I found that on several other timelines or dimensions, we really did not wake up and get out of the house in time.

That we had both died from carbon monoxide poisoning. This was quite startling and yet very interesting.

However, using muscle testing once again, I found that on other timeline dimensions including the one that I am writing this exact book on, *we did NOT die and survived.*

Multiple Living Versions Of You

This Theory of Reality explains why, after you experience some personal improvements such as meditation, mantra chanting, intense prayer followed by a total acceptance of your situation the people close to you quite suddenly become so much nicer and everyone becomes more pleasant.

No, they really did not change at all.

Instead, your own conscious self's vibration has aligned you to a different timeline reality and dimension where the people close to you are nice all the time. You can try to stay there but reality is constantly changing so you probably won't. The world is invisible waves of energy until we manifest it by observing and experiencing, so therefore nothing is actually real until you start paying attention to it. So the people are actually a newer version of themselves being experienced by you.

IT'S THE MOLECULES SILLY

It's because of the molecules that make up everything in our world.

Molecules are constantly in motion, and at the root of everything we see.

What we learn from quantum physics is that nothing in our world exists until we create it – at the very moment we observe or experience it, those molecules start to react to us.

So once again, why don't more people notice when these physical changes happen?

We oftentimes put too much trust in our basics senses without seeing if there is more than meets the eye. This is because often doing so creates simple answers to otherwise very complex questions.

If you want to experience change in your reality, the safest way is to let the universe do it. So that means you have to accept the reality of where you are now.

You have to be at peace with the process.

Your part of this is choosing that the journey will go smoothly, asking for help so that it would be wonderful and accepting that the ease of the trip is in your acceptance of what already is.

You see this is the way that quantum physics has everything to do with every single part of what exists in our physical reality. Scientist have already discovered through mathematical equations and actual experiments that particles making up matter in the form of a flower, chair, human, or an animal are literally "projections of a higher-dimensional reality which cannot be accounted for in terms of any force or interaction between them"

The subconscious mind is beyond your normal awareness, so you have to make special efforts to access it. In fact, until you observe or experience your reality, it exists as invisible waves of energy.

Gratitude is one of these beautiful waves of energy that comes from the heart. It's genuine. When you honestly value what you have, the gratitude energy becomes concentrated and reality changes occur. While many people embrace these ideas, there are quite a few who either dismiss them out of hand or else make it clear they find these ideas frightening, or at least a challenge to what they think is correct.

It is scary to change beliefs you've held for a lifetime. It is almost unbearable to change a lifetime of assumptions from other people, and have to start thinking for yourself.

Lately, I have found myself frequently expressing to people that, while I respect their opinions and beliefs and am glad they have found what works for them in their lives, I believe differently and their views do not work for me. What is "right" is what is right for them. What is "right" is also what is right for me.

Reality depends on what actually happens (objective) and how our brains make sense of what happens (subjective).

These are the necessary components of reality, and reality is a subjective concept unique to each of us.

Experiments have determined that subatomic particles, which comprise all matter, are not even solid, stable objects. They are vibrating, indeterminate pockets of energy that cannot be understood or defined in isolation.

They are dual in nature, sometimes behaving like a wave and sometimes like a particle and sometimes even behaving like both AT THE SAME TIME.

They only "collapse" into a set state in the moment they are being observed.

So our acceptance of situations in our life is being aware of awareness. It is approaching the present experience with an awareness including the qualities of openness, acceptance, and love.

If we struggle with or constantly resist change, we will always be behind the wall trying to regain something that doesn't exist anymore. Something that stopped collapsing the moment we stopped observing it, and became wave or particle energy.

In other words the old things don't really exist as they did before.

I have always believed in the golden thread that ties seemingly unrelated events together, and I always trusted in the qualities of crystals, herbs and flowers.

Somehow I have always known that simply putting your attention (and intention) on something boosts its power in your life.

<u>So you have to practice being in the NOW, if you want your life to change for the better</u>.

Being in the moment is about putting your phone away and really looking around you as you sit at home. It's about making eye contact with your friends and really listening to what they have to say, not just waiting for your turn to talk about yourself.

Being in the moment is about removing all the distractions around you, closing your eyes and tuning in to you.

After many time shift experiences, I no longer feel that time behaves in a linear fashion.

I do not believe that time always moves at the same rate either and I'm not the only one who feels this way.

Yet, most people never give much thought to the fact that there are different layers of reality depending on your awareness.

Time Loops in which a person enters a place twice in exactly the same way, and when questioned, asserts that this (the second time) is actually the first time they have entered the place.

Time Slows to a Stop in which you are moved from one place to another place in a moment, with no recollection of how or any loss in time.

Retroactive Prayer in which a person invokes their higher power to alter past events in their memory in order to influence the present or the future.

These are just a few of the many layers of reality, which is virtually unlimited and there are no borders between this and all other dimensions.

Reality is alive and changing all the time. More importantly, our consciousness is a part of this picture demonstrating that parallel universes exist and can be reached.

It took me a couple of months of prayer and meditation to begin to re-wire my belief system. I feel like I'm gaining a back-stage view of how the universe really operates.

Conscious thought has been demonstrated to change the structure of water and remove pollution from the skies and incredibly it's been linked to miraculous changes in our DNA.

Dimensional shifts happen only when you've let go either intentionally or accidentally.

If you are struggling with something difficult in your own reality and you are very stressed out with this struggle, a dimensional shift will not happen.

You first must accept that you are where you are, and just stop fighting it, stop trying to go against it.

Allow yourself to just step back and let everything flow freely.

You must just relax your mind, in order to give in to your situation. Usually this means that you've accepted where you are, you are thankful for where you are and you just let go.

Stop telling everyone how much you don't like the situation you are in, and start saying that you are thankful and grateful for where you are right now.

Once you let go, you release the hold and you are ready to be transported to your next reality. Dimensional reality changes are usually triggered when you are happy and feeling love.

We are all in the process of moving from one dimension into another because it is a lifelong process of progressing toward ascension. However you can slow down the process with resistance.

You cannot move from the dimension you are in to another dimension until you have the right energy to stay there. This is because you gradually shift to higher-vibrational realities.

This is after struggling and resistance falls away from you and your responses to the realities you experience becomes one of acceptance and gratitude.

When you are able to see the blessings and higher order contained within every situation, new thoughts will filter through your mind of possible alternatives and choices.

Changing your reality does not change who you are. However, reality changing shifts can alter your beliefs, perspectives and sense of possibility.

This can in actually change your past, present and future and allow you to be in tune with new and different dimensions.

Accepting a multidimensional reality is one of the most significant concepts that you can grasp in your lifetime.

The way you look at everything changes.

You receive the ability to experience reality in a larger field than you did before since you are more aware of there being more to life than you previously believed.

In quantum mechanics, there exist several theories that include the existence of parallel universes.

The Copenhagen interpretation of quantum mechanics states that an object exists in all of its possible states at once (coherent superposition). It is only when we observe the object that it chooses a single state which is of course the one that we observe.

This is demonstrated by the famous thought experiment, "Schrödinger's cat," as well as the Double-Slit experiment.

The Copenhagen theory is dependent on the observer which would be you.

The Many-Worlds/ Dimensions Theory states that, for every single possible outcome to any action, that world or reality will literally split into another copy of itself.

This is known as DE cohesion.

Therefore, each possible outcome will exist in its own dimension or world, so that every possible outcome will occur.

Therefore, unlike the Copenhagen theory, which is dependent on the person watching, the Many Worlds Dimension Theory is independent of the person watching and is controlled only by the universe. The world as it is and the world as you see it are two very different things.

Did my daughter and I die only to wake up in another dimension and timeline?

It seems more than a feeling - yet I can't say for certain and therefore I must consider all possibilities.

So that's exactly how I explain it... the only way I can explain it.

I feel like I died, and then I woke up in my bed.

There was no separation of time between the "events." I remember the day of the incident and all of the events that led up to it.

I remember waking up as if time slowed down, probably due to anxiety. So to me after dying,

I woke up with my daughter coming out of her room at the same time - only somewhere different in feeling. The house looked the same but felt quite different.

I remember things swaying. Everything around me became distorted and I felt as though I were in a fog, or a dream, wandering out of the house.

As far as my mind is concerned, the incident happened, and everything changed after that. Then the world felt different, somehow and many physical things seemed altered from what I remembered them being the day before. I feel like we were moved to another reality dimension to complete "unfinished business".

Whatever it was, it definitely has continued to reside in my mind.

It has stuck with me enough to lead me on a journey to find the answers.

It has also become real enough for me to feel extremely uneasy every single time I drive by or approach the community where we used to reside.

Perhaps this was one of those reality shifts - as a gap in reality seemed to form, become foggy and distorting things temporarily perhaps merging two dimensions for a few moments... and then close.

So that in this reality...the one where we did not die, we simply walked out of the house.

Through the practice of meditation I have felt the true essence of my being and my interconnectedness with the universe and everyone around me. It has become apparent to me recently that I am guided now more by my inner self than I am by my thoughts.

According to Albert Einstein, there are four dimensions, three of space and one of time, with the special property of the ability to bend light.

Stephen Hawking furthered this theory by questioning the possibility of many more dimensions existing in the realm of our universe. With so many scientific ideas being explored today it makes the idea of dimensional changes so much more than a great possibility.

Chapter Two

The Gratitude List

It is no accident that you have started reading this book. There are no coincidences! Are you in any one of the following situations…?

Reliving your old circumstances over and over again?

Engaged in repeated patterns of misfortune, trauma and negative situations?

Feeling frustrated about being unable to find a pathway to your

imagined life?

Feeling stuck despite having tried all ways to create a change in your life?

Are you always lacking in some major life aspect?

Is making money a major struggle for you?

Here is a very revealing question.

Have you ever really been grateful?

To begin with, you have been looking in all the wrong places. If you woke up this morning with more health than sickness, you are more blessed than the millions who will not survive the coming week.

If you can read these words then you are more blessed than over two billion that cannot read anything at all.

Do you practice gratitude for all of the blessings that you do have. The universe has some very real laws.

The by- product of one of them is that the more you focus on what you do have the more it will become available to you.

Close your eyes and see it there.

 Be silent and have a taste of it. Your very nature is to be blessed and to be grateful. Yet, when you meditate on this thought and begin knowing it as the Truth, you will discover something that turns your entire thinking around.

Nothing ever happens by accident in this Universe.

There are no coincidences. Life is the perfect teacher.

Every lesson your soul came here to learn is being provided for you through each moment and experience that you are given.

The paradox is that we are to be grateful not just for the things we think of as "good," "fortunate," and "successful." Not just for our peaceful times and joys. No, from the perspective of the universe, we should be grateful for ALL our experiences. As we stop resisting and are able to say a heartfelt "thank you" for this adventure in life; life itself starts to transform for you.

Make it a daily habit to express your gratitude energy and appreciation energy for the wonderful things and people in your life. Be thankful for your loved ones and friends, your job, your pets, and anything else you have in your life.

It's not just that we feel grateful, or that we express our energy in the form of gratitude, but that we actually experience a sincere desire to give something back. Think of it as appreciation energy that promotes a sense of obligation. Not an externally forced obligation, but a thankfulness that arises naturally when you understand that the Universe has supported and cared for you.

Be warned: It WILL change your life, but it will do it in ways you cannot imagine and bring you such a great taste of life you cannot even begin to initially comprehend.

Imagine that you are walking on a street and a brick wall has been placed directly in front of you. It's solid, it's real, and it's obvious that for you to continue this wall has to be removed.

Alternatively, you can to choose stay exactly where you are and have this wall stop you from walking anywhere. Perhaps your subconscious has been trained to believe that the wall will never be removed. Until you believe that you are able to move past any wall, you won't be able to move past this wall.

It starts with learning not to complain. I find that impatience goes hand in hand with complaining. You have the opportunity to be patient with the rest of your entire life; because every moment has something of interest if only you will slow down and see it, and stop trying to be the dictator of your life.

If we cultivate a mindset of gratitude, even for one moment in a situation where we could find ourselves complaining, the healing reaches into our minds, and will eventually radiate into anyone you come into contact with.

There's science behind this which suggests that complaining and cultivating negative states can be detrimental for our well-being, and a mental awareness such as gratitude actually boost our immunity. There's that gratitude energy again!

Did you know sharing gratitude with someone has been shown to increase happiness by up to 19%. Further research suggests more hope, optimism, a better ability to manage stress, a tendency to exercise more and the benefits of sleeping better.

Those who have a positive outlook on life perform better than they would if they had negative thoughts.

In fact your brain performs 31% better when it's positive than it would if it was negative, neutral or stressed.

This will please the Lord more than an ox or a bull with horns and hoofs; Let the oppressed see it and be glad. You, who seek God, let your hearts revive.

Psalms 69: 30-32

Since not all of us are as naturally inclined toward feeling grateful, the research also suggests there are intentional steps you can take daily to turn your mind toward being truly grateful.

Simply writing down the things you're thankful for, on a regular basis, seems to bring on many of these benefits. In has been determined that at least 90% of your long-term happiness is predicted not by the external world but by the way your brain sees and accepts the world.

So now back to the brick wall directly in your path.

You may be without a home, a job, dealing with an illness or facing another difficult situation of some kind. This may be the form you're your brick wall has taken. Your job is to determine that no matter what you are facing, you will make your mind up to remain grateful, and have a thankful attitude. In other words you are using the force of this energy to get through the wall.

This is extremely hard to do when you feel like the world has crumbled and fallen in on you. Yet, that is when you need to do it most.

During the most stressful times in my life, I make it a practice to find a few minutes every day to say out loud, "I thank you for…"

I start with being grateful for various people. To me, that's the most important thing.

If I remember to be thankful for those people who are supportive of me, I feel as if everything then were a miracle, and being aware on a continuous basis of the miraculous opens the doors to greater things.

After that I go through the "material things and experiences in living," that I'm thankful for.

That brick wall appearing in your path carries within it a great benefit. In the face of this adversity ask yourself: 'What's good about this?", "What can I take away for my life from this?", and "How can I learn from this?"

You thoughts and feelings create forms of subtle matter which are of a substance every bit as real as any material substance. So this wall is a reality that must move in order for you to continue on your life path.

It creates depression, anxiety, and self-sabotage. Because it blocks your abilities, it will also block you from success. The physical effects of this wall in your path, are often pain in the neck, discomfort in the shoulders, or tightness in the shoulder muscles.

This feeling of pressure or discomfort in the chest and neck occurs when we feel attacked, grief, deep sadness or loneliness. I have experienced much grief, sadness, loss and heartache. I know and understand first hand that emotional pain can and often will manifests in the physical.

Anything you feed energy to, will grow. So, it's really important to look at what you're feeding with your energy.

We need to release that which no longer serves us, we need to look at what our bodies are trying to tell us, and we also need to learn the lesson of accepting our circumstances and learning from everything. That is why the energy of appreciation and gratitude are so powerful.

I began studying quantum physics because I was curious. I became enthralled when I realized the similarities to how faith works.

It took faith to believe that you can have what you say.

QUANTUM PHYSICS AND YOU

Quantum physics is the study of things so small that we cannot see them, yet everything we see is made of these subatomic particles. Words and thoughts are energy and energy affects matter. Those words and thoughts are vibrations of energy that affect the atoms that make up the things and events of your life.

Scientists have performed experiments with atoms and their subatomic particles such as electrons.

This is because the electron that is seen orbiting the nucleus is not always there in particle form. It exists in a wave state (sort of like a cloud, everywhere at the same time) until someone looks at it.

When the scientist begins to look at it, it suddenly appears as a dot (particle). What is amazing is, "How does it know someone is looking at it?"

It is responding to the observer's interaction with it.

What??

One of the difficulties in quantum physics is that the particles behave uniquely for each observer, so you may well ask "Does it behave according to what different scientist may believe?"

Actually yes.

This is because the thoughts and beliefs that you carry also produces energy around you. Everything has a frequency of vibration. Even gratitude and appreciation has this frequency of vibration.

Quantum physics is an area of science where the known laws of physics (Newtonian physics) no longer apply. In classic (Newtonian) physics you can repeat experiments using the same formulas and get the answers and responses you expect.

This does not happen in quantum physics.

In the quantum, subatomic area, there are only possibilities and probabilities. This is because as we already know things don't work like you thought they should. Nothing is there until you look at it or give your energy and attention to it. All that exists is only an infinite number of possibilities.

All things are made up of atoms. Atoms are made of subatomic particles. These particles are not really particles because they exist only in a pool of possibilities. They are just an idea until someone observes them at which point they appear as a thing (particle).

This is why when you believe and pray for something it actually only exists in your mind and is only a possibility. By placing your energy on the thought it becomes a probability.

There are an infinite number of possibilities that exist for your life. It is necessary to let go of old ideas and open yourself to new ones. You must let go of your old beliefs in the way you think things work.

In quantum mechanics, the observation of something changes it. You cannot be certain if it existed before you looked at it.

When you looked at it, you actually interfered with whatever it was before you looked at it. If this sounds like circular reasoning I guess on some level it is. It just means that we affect everything around us just by how we see it or what we believe, and what we believe affects everything around us.

This is why gratitude and writing things down affects everything around you. That is why it is so important to change your perceptions, beliefs and expectations. What you believe in your heart is the controlling energy for your life.

In quantum mechanics, the many-worlds interpretation suggests that every seemingly random quantum event with a non-zero probability actually occurs in all possible ways in different "worlds", so that history is constantly branching into different alternatives.

In *The Greatest Gift: A Christmas Tale*; a book by American author, and noted Civil War historian Philip Van Doren Stern, on which the popular movie- *It's A Wonderful Life*, made in 1946 was based we find just such a history branching scenario.

George Bailey got to see what the town and his family would be without him. The best thing about this movie is the realization that time itself can be transcended, that it is possible "to live in the past, present, and the future" simultaneously.

Here you read of a man who tries to kill himself yet inexplicably is shown that he lived each moment having created an alternate, reality, a new timeline to a new and better future, than the one he thought would be better without him. He sees a world that never was, a world without him, an alternate dimension.

He sees the consequences of actions, by the absence of those actions. As Clarence says, "Strange, isn't it, George? Each man's life touches so many other lives. When he isn't around, it leaves an awful hole." These questions are actually demonstrated in this popular classic film.

"How do our actions affect the future? What impact do I have on the universe? Can I define reality for myself? Do I have a destiny?"

Just like George Bailey, you can choose to live again. When he realizes that the world he knew of "Bedford Falls" has fallen apart as a result of his choice. He reverses his decision, and emerges back into the original future with a renewed appreciation for the effect he had on the world and the people familiar in his world.

I have watched this movie at least a hundred times, and it wasn't until I started on a person quest for answers for myself, that I even realized the implications of the angel, God and parallel dimensions where I don't exist, as shown in the film.

Time and time again George is forced to put off his dreams until he gives up on them completely. He doesn't value his life or his achievements, going as far as to wish he'd never been born.

This eventually leads George to jumping off of a bridge; not in suicidal desperation as he originally intended, but to save a drowning man. He instinctively jumps in after him, forgetting for a moment that he had been contemplating killing himself just seconds before. They are both pulled from the water by the tollhouse keeper.

The drowning man turns out to be his guardian angel, uses this tactic to get George to stop thinking about himself and to try to save someone else.

It's a Wonderful Life's story is an alternate dimension that was accessed with an angel guide in which George never was born.

The concept of sideways time travel is often used to allow characters to pass through many different alternate histories, all descending from some common branch point on a persons' timeline. Often worlds that are similar to each other are considered closer to each other in terms of this sideways travel.

This literary interpretation is sometimes inspired in the many worlds interpretation of quantum mechanics formulated by the physicist Hugh Everett, as an alternative to the Copenhagen interpretation originally formulated by Niels Bohr and Werner Heisenberg.

So therefore a parallel universe is a permanent world, universe, or dimension that exists simultaneously with, but separate from, our own.

In most cases the parallel universe can be entered by characters from ours, or vice versa. This universe exists in physical space.

In *The Universe in a Nutshell (2001)*, Stephen Hawking writes of a sports multiverse, declaring it "scientific fact" that there exist a parallel universe in which Belize won every gold medal at the Olympic Games.

Theories don't prove that something exist, they prove that under the right conditions, something is possible.

Surprisingly, however, the idea of parallel universes is far older than any of these references, cropping up in philosophy and literature since ancient times.

This concept is found in ancient Hindu texts, in writings such as the *Puranas*, which expressed an infinite number of universes, each with its own gods.

In religious ideology our human existence makes possible the multiverse theory. Since it's not possible for the Universe to have come from nothing and since we also exist; it means we have always existed, even if only in the mind of God.

When it's a decision between a finite or infinite God, then there is no choice. If He can do it once, He can do it an infinite number of times. It's all about what is possible.

The entirety of space, time, matter and energy is all happening at once in different timelines; it is parallel universes.

Maybe there's a universe where you get the life you want.

Where you don't second guess everything and are not afraid of change.

Imagine a universe where you actually end up with everyone who appreciates you. Maybe there is one where no one is treated poorly, or taken advantage of.

If this theory holds, well, by the law of averages, there had to be one universe — just this one — where things don't end well.

Here and now just happens to be it. If you think of it this way, nothing is our fault, or a mistake.

Could ingratitude trigger dimension changes causing things of which we are unaware?

So now back to another brick wall.

What would you do if you found yourself in a strange bed, your office and job changed, and you discovered your boyfriend no longer existed? I recently read of one such example.

Lerina García claims this is exactly what's happened to her. Could you be next…?

What started as an ordinary day just waking in bed one morning turned into a series of shocks and horror for a desperate woman lost in a strange and different world: our dimension.

First, she noticed the sheets and bedclothes were strange. She didn't recognize them at all. She had entered another reality. Her life, her past and everything most precious to her was gone.

As she went through the routines of her first day in another dimension she noticed small things out of place, items missing or things she hadn't bought in her home surrounding her.

She said "One day I woke up and found that everything was different—nothing spectacular or having to do with time travel and such things. I simply woke up in the same year and day on which I went to bed, but many things were different."

"They were small things, but sufficiently important to know that there was a point at which everything was different."

Not everything was small. Her car seemed the same. She still worked at the same company she had for 20 years. Although in the same building, she was shocked to learn her department no longer included her. Her office was now in another department in a completely different part of the building.

She said "I still worked there, but in another department, reporting to a superior I didn't even know." "I went to the doctor and underwent drug and alcohol testing…all clean." "I returned to work the next day and was able to make my way by asking questions and saying that I wasn't feeling well."

"I've been separated from my partner of seven years for some six months. We broke up and I started a relationship with a fellow from my neighborhood. I know him perfectly well, having been with him for four months. I know his name, surname, address, where he works, his son from another relationship, and where he studies."

"Well, that fellow no longer exists. I've hired a detective to find him and he does not exist. I've visited a psychiatrist and it has all been put down to stress. He thinks they're hallucinations, but I know this isn't the case. My former boyfriend is with me as though nothing had happened—apparently we never broke it off [in this world]—and Augustín (my current boyfriend) appears to have never existed. He doesn't live in the apartment he used to live at and I cannot find his son."

"My own family doesn't remember things like surgery performed on my sister's shoulder a few months ago: she says she's never been operated on, small things to that effect."

"For five months I've been reading all of the theories I've come across and am convinced that it has been a jump between dimensions, planes or something, a decision or action taken that has caused things to change."

Could Lerina, a highly educated woman be simply hallucinating and imagining everything? Perhaps she is suffering from a form of rare time-related mental illness?

What about the other poor woman she changed places with on the other dimension? Could she be mentally disturbed as well?

<u>Perhaps not since she's not alone in her experience</u>.

Then I learned of another such incident.

ANOTHER DIMENSION?

A curious incident took place in Tokyo, Japan during the early 1990s: a man arrived on a flight with a passport from a non-existent country.

The man expressed anger and shock when Japanese customs officials detained him. Although the officials checked their records carefully, the passport had been issued by a country that did not exist. No record showed the country had ever existed.

Although passports exist issued by non-existent countries (known as camouflage passports), this passport was real and had custom officials' stamps on various pages including stamps by Japanese customs officials from previous visits.

The man was well-traveled, Caucasian, said the country was in Europe and had existed for almost 1,000 years. He carried legal currency from several European countries, an international driver license and spoke several languages.

Finally, indignant, he demanded a meeting with higher government authorities. He was convinced some massive practical joke was being played on him.

After being detained for almost 14 hours in a small security room at the airport terminal, some government officials took pity on him and transported him to a hotel. They ordered the mystery visitor to wait there until they decided what to do about the matter. From the reports, the Japanese were just as confused and flustered as the mysterious man without a country.

Although two immigration officials were posted with instructions not to permit the man to leave his room, the next morning the guards discovered he was gone.

The only exit was the door they watched and the only window had no outside ledge and was 15 stories above a busy downtown street.

The authorities launched an intensive manhunt throughout Tokyo for the mysterious traveler, but finally gave up the hunt.

The man was never seen again.

There is no scientific explanation for any of the incidents described, unless one looks at the leading edge of scientific study on the multiverses.

In the world of the quanta—which encompasses all that is; the multiverses are vibrating differently and some parallel worlds literally are intersecting with each other.

Physicists also have had glimpses into these other realties during experiments where sub-atomic particles have winked out of existence and then reappeared.

Could a person slip from this reality into one "next door" to our universe? If so, would there be an exchange such as trading one version of a person for another?

Time would be unaffected, and the two parallel universes would be so symmetrical that only minor things may be different.

Whatever we may think and whatever the truth really is, I do believe that these people who have had these experiences are sincerely sharing something that is very real to them. What is reality? It's easy to pass judgment on others by thinking that our personal reality is the only one there is. Considering the secret experiments that have been and are still being conducted here on Earth, in this dimension we should not be surprised that the unexplained has been happening. The military had been tampering with time portals since before World War 2 and something seemed to have gone wrong. They changed something in the past. Something they did not intend to change has brought about severe dimensional alterations.

Many people have had the experience of looking for a missing object then finding it a day or so later, in a spot you had checked many times before? Maybe seeing something that was so out of place that you had to rub your eyes in amazement?

If we exist over a wide range of universes/dimensions in the multiverse do we exist as separate consciousness or is there a universal element of this connecting to all of many ourselves?

Are You On The List?

I started recalling recently that I was always noticing odd little things happening and/or remembering them "differently."

There's a number of things I remember about my childhood and early adult life that I recall with pin point clarity.

 Yet I was always challenged about my memories. I was in my early 30's, and I would get so angry with my friend saying everything was so different from what I just knew it to have been.

I always found these conversations puzzling. How could she be so wrong about so many events that I remembered so well?

I still suspect that the truth is the same for each of us, because we are both different now. There are many people who don't feel "right" where they are in this world dimension and know something feels different. Could we both have been correct?

Shared memories are totally different now with various people in their past, while familiar places aren't even oriented to look the same anymore.

So what does this have to do with my wall?

The good news is that the Most High God still transcends time, multi-verses, dimensions and walls.

He is has placed the wall there for a very good reason.

God gives you exactly what you want and watches to see how you journey back into acknowledgement of the universal consciousness in you for being grateful.

If you think that everything is just an illusion in your mind, you see that the only thing of real value is your treatment of the beings/people that are in your life.

Open your mind to another possibility.

That what you perceive as an obstacle is actually something completely different.

There's a lot more going in this world and in this universe than we can possibly imagine. The way you recall events will change as you and the Universe changes.

Gratitude

When you consider all the many things being experienced by other people in the world, just how grateful are you for your life?

Express gratitude and count the good events you experience each and every day. Start by making a list of all the things to be grateful for in your life.

Remember, just like the fictional character George Bailey when you don't value your life or your achievements, you open the door to dimensional changes that operate in a lower sphere from perhaps where you really want to be. Since those lower energy thoughts and feelings are full of ungrateful and negative emotions you'll most likely attract negative things, situations and people into your life or dimension.

Say "thank you".

Really how hard is that.

You can say it in person, over the phone, in a note, or in an email. Using this when it's least expected can have significant results in altering your entire life.

You see as people, more money is only one of the motivations that may drive us. We also look for respect, a sense of meaning and a sense of accomplishment. Saying "Thank you" well, it really doesn't cost anything, but it has a measurable positive impact on you the recipient and as it now turns out, the universe.

Stop complaining about your life.

This may come as a surprise to you but nobody really wants to hear your complaints.

I'm tired of people complaining about things they can't change, but more frustratingly, the things they can and refuse to change.

So just how do I know nobody wants to hear?

Because I was one of those people. I complained until I was exhausted, and of course until people stopped listening to me.

No one wants to hear the same negative information in a different package every day.

Life is too short to spend time with people who suck the happiness out of you with ingratitude. *You see complaining is ingratitude.*

Well, perhaps it doesn't sound like much of a problem at first, but it can be very serious.

It can keep you from getting to an enjoyable life. It can keep you from having gratifying relationships.

People don't want to "keep company" as my grandmother would say with someone who complains all the time.

You cannot be grateful and complain at the same time, because gratitude is a lifestyle mindset choice that will drastically change your life and your environment. This type of thinking is powerful because it causes us to be more mindful before we speak. To actually consider if what we are about to say has any value or is silence so much better.

When you're grateful, you tend to exude and share that contagious positive energy. When you start appreciating your life, people around you start appreciating you.

Be grateful for having seen the beauty of a flower, the sound of rain, even if it happens as you are getting drenched caught in a storm. Just be grateful for that experience, and place it on your list.

See the brick wall differently.

Generalized gratitude is the attitude of the appreciation of all things in your life and in the world. The universe has a gratitude list as well.

So are you on the list?

Is someone somewhere grateful for you?

This idea is simple but incredibly powerful. It's about taking the time to notice the good things and people in our lives and getting more of them daily.

You were sent here with purpose, is it gratitude?

Every day I am amazed by the ways in which I am was so richly and undeservedly treated with respect, kindness and thoughtfulness.

The person who showed me friendship, and kindness by complimenting my old scarf, is on the list.

The Lady who noticed my manicure and even laughed at my old jokes, is on the list.

To those people who have held open doors, delayed elevators so I could share the ride with you, or carried my packages to my car; I am so grateful for you and yes, you are also on the list.

You gave me the feeling that you were happy to be able to help me, and then went on your way. Believe me your smile made my day a whole lot easier.

To the often nameless people who did personally make my life better, I can be grateful for you and put you on my gratitude list.

I know you were sent to make my day better. I'm sure you did the same for someone else and are on their list as well.

Every night - before you go to bed, think back over your day and remember three good things that happened.

Then allow the Universe to fill your life with Gratitude, Love and Acceptance, knowing that the wall is in your life for a reason.

Chapter Three

Living One Day At A Time

HOW DO YOU FEEL TODAY?

Take a minute each day to listen to yourself.

Shut out the noise, close your eyes, and allow yourself to understand how you truly feel right now.

It is by understanding your true feelings that you will know exactly where you are and what to ask for specifically.

The power of the Universe can add a great deal of support to your desired thoughts, because this Universe is a matrix of every ingredient.

It holds our experiences, relationships, environments, things, concepts, memories, thoughts, dreams, emotions and everything else that makes us who we are.

But really who are we?

Are we different people at different times, or just facets of the same being?

At all times we have at our disposal the vast resources of our Universe. Yet we may not really understand what that means.

Our mind wants to make plans, worry, stress, and try to figure out how it's going to solve our problems, but little do we know that all that is actually PUSHING the solutions farther and farther away!

What if our ideas are based on not understanding?

What if it is all just a thought away ?

Sometimes when I want to make a significant change in my life, I think of it in terms of changing dimensions.

My attention is centered on my current reality right now, and my desired situation can be said to exist as some alternate reality outside of my concentrated focus.

In that other reality there's another me who's already where I want to be. My goal then is to become that other me and to shift and merge into her reality.

There is an infinite layer of possible universes.

Know that in fact they are all YOU.

You exist in multitudes, but you normally focus your attention on just one of all possible realities.

Nobel prize-winning Danish physicist Niels Bohr explained: "Isolated material particles are abstractions, their properties being definable and observable only through their interaction with other systems."

According to another noted physicist, matter cannot be objectively perceived or described apart from the observer – matter and mind are co-dependent.

In other words the observer is effectively a participant in the reality being observed.

When the mind begins to recognize it would like things to be different, a huge opportunity has been born.

Einstein said, "It is very possible that in a different dimension right now, a railroad train is coming right through the middle of this room, only we can't perceive it because it's in an entirely different dimension."

In the simplest sense, a dimension is a band of energy frequencies that create a "world."

There are infinite dimensions in the Universe, just as there are infinite frequencies of energy.

So your many thoughts are energy, and that energy creates "many worlds."

SPIRITUAL PHYSICS ??

I had been pursuing the nature of the structure of reality for decades, most especially in recent years. Physics, like spiritual philosophy, is engaged in a search for the nature of the universe, so it was not unexpected that both pursuits would eventually collide.

The physical conception of these extra dimensions was (and still is) the hardest part of the theory to comprehend. The spiritual concept however is easier to my mind.

According to quantum theory, a domain of reality exists, which does not consist of material things but of ideal forms.

These ideal forms can exist into infinity and can also be infinite in number.

The multi- dimensional versions of you…….

For example, if you walk out of your front door, and see that you can go right or left, the present universe gives birth to two more universes: one in which you walk off having turned right, and one in which you walk off having turned left.

So that in each universe, there's a copy of you envisioning one or the other outcome, thinking very incorrectly that your reality is the only reality.

If God could then be an unlimited entity, this would be reflected in His Universes which would also be unlimited. There can be infinitely many different particles and so they can be arranged infinitely.

What if there are nothing but right choices and right decisions?

Then you are always in the right place, making the right decisions and therefore heading to the right direction.

So why be worried ?

Hmmm……..

So even the desire to go to another dimension is a right desire!

If you think something is "possible", you can calibrate your consciousness to the frequency of reality in which it naturally occurs.

Yet it all comes back to you.

<u>*Can your mind change your location ?*</u>

When things seem tough, stop for a moment and think about your place in this Universe – about the total abundance and richness around you. Think how many stars there are, how many fish there are in the seas, and how many grains of sand on the beaches.

Think of how big the ocean is; of the immensity of space, and the bigness of everything. How it is all connected….

Have you ever been amazed at how something that seemed impossible "just worked out" or how you managed to overcome another wall? This is because there are powers in the universe that we have no idea of, yet we have access to so much help if we only just ask for it.

There is a natural law termed Reciprocal Action, which guarantees that whatever you are thinking about, speaking about and concentrating your mental energies on must return back to you in either a spiritual or physical substance.

Everyone is doing this, twenty-four hours a day, whether they know it or not. You are the words and thoughts you have been having.

You have packed your own bags.

All disappointments and failures are the result of energy being drained away from what we say we really want and applied to produce what we have really been thinking about.

In the Bible the story of Daniel has always fascinated me.

Daniel was thrown into a lion den, to the hungry lions he was

just food, and I believe Daniel knew this as well.

If you were thrown in a lion den, no matter what else was going on, your new number one problem would be Lions.

You would talk about lions, think about lions and pray about lions.

However Daniel did something that inspires me still.

Daniel prayed to the Most High, <u>and then turned his back on the lions!!</u>

Daniel showed by this, that as far as he was concerned it was already morning and he was already safe and delivered out of there.

He had no reason to believe that it could happen, and every reason to give up. But instead, he chose to think positively, and KNOW that he would be safe and out of the Lion Den in the morning.

Yet, how did he get in there?

Scripture shows that he was worried about ending up there to begin with.

Yet he quickly figured it out, because his life was on the line.

In order to exercise faith, there must be a removal of fear.

He had to put all of his awareness on walking out of that place in the morning and nothing else.

He had to have a conscious faith.

<u>In order to walk through your wall, you have to replace fear with the solution and nothing else.</u>

Walls are what turn simple things we want to do into difficult things. The ability to completely remove from your mind the awareness of the negative problems, deprives them of any energy in their formation, or power over your mental state, and they will dissolve from your life.

It's like walking through a wall that's floating in the middle of nowhere and all you have to do is feel for the opening and let yourself float right through it.

You can't really force yourself through it. You just have to float. You have to let go and trust the Universe.

The challenges we face today in society and in our personal lives can be daunting. Sometimes you have to learn how to trust even yourself again, trust your heart instinct, trust in the voice inside that tells you "you're going to be alright".

As you succeed in moving through a wall, you build the confidence to take the next step. You don't need to do it quickly, and you don't have to do it perfectly.

The very first thing I did was say; "you will get through this and you are going to be alright."

Just like Daniel I turned my thoughts to my desired goal and not on the problem. I had to learn quickly to replace the fear in my heart, with love; not blame, not anger, just love.

I had to think on the good continuously, and nothing else.

Like Daniel I learned that I did not need to worry about surviving the rest of the year or even the month I just needed to get through TODAY.

You can never realize your desires if you do not know positively and in detail what you want and when you want it.

I think a lot of the time the confusion is that we don't actually know what we want to do/receive/be and so just getting to that point is a major step.

Each word you speak or write possesses its own vibration which is felt inside your body the instant it is said. When you say something out loud, or write it down you are in actuality proclaiming who you are and or who you will become to the entire Universe.

In silence, you become aware that you have the freedom and power to choose the types of thoughts you wish to entertain and empower and the words you wish to speak.

It is when you are silent that you are expressing your faith and trust in the Universe and in yourself.

Amazing, incredulous things seem to happen to me all the time. I still call them wonders or just miracles from God.

The amazing thing though is that when I am specific with my requests, what I want usually happens exactly as I envisioned it. Strange little synchronicities were popping up everywhere.

I couldn't ignore all the miracles happening in my life.

 I was being swarmed.

 There was no way that I could ignore these synchronicities.

But it's more than that.

You have to open a conversation with the Universe that is going to taking into consideration your request and to do this you have to be positive and give up trying to have control.

Letting go and letting life move through you sounds like weakness, but going through life without a fight requires more courage than kicking and screaming at every turn.

When I would say something like, "I'm tired", to which my daughter would often reply,

 "You're not tired, you just need to relax".

She sounded as if she thought that I didn't know how to relax.

Well I was always busy, because I was always mentally working on – either trying to solve problems, learning something new or planning to avoid future problems.

She was right.

I didn't know how to relax anymore.

I couldn't sit still and therefore I couldn't relax.

I realized something had to change.

However it is hard to learn to change and even harder to learn to relax.

Yet I know and understand that to truly relax you have to clear your mind and do absolutely nothing.

Setting aside some free time to do nothing on a regular basis is very healthy for your mind, body, and emotional life, especially if you find that you're really wearing yourself thin.

Doing nothing, just sitting and thinking, is very restorative, and terribly undervalued.

Clearing your mind and learning to relax paves the way to learn to meditate and hear from the universe.

This is the best time to speak to the Most High in your thoughts.

Choosing to be true to you is tough, and it can be risky because you will learn some disagreeable things about yourself. You can also speed up your release during difficult times by looking within, seeing the beauty and lessons of your current experiences.

This will reveal your true identity, your faith and trust are seen by being silent, waiting and watching for your miracles to appear.

The greatness of any success you have will be directly proportional to the greatness of your thinking and belief.

Your thinking and your beliefs about your thoughts determine your feelings and your actions.

Those actions are based on those feelings.

"Knowing" and "Believing" are two different concepts.

You can believe that things will one day change for the better, and you can continue to hope.

However "knowing" is having the understanding that things already have changed for the better, because it was predestined, ordained and is an irrefutable law and fact of the Universe established for you. Knowing is NOW.

The most interesting thing is that the path to great achievement is usually started down on by some major catastrophe or crisis occurring in the life of a person.

This is when most people discover that there really are two of them in every existence, and the self that cannot be defeated begins to emerge.

Multiple crises are opportunities for multiple successes, or multiple failures.

Focusing on the positive aspects of every seemingly negative event, can immediately "flip the script" and begin to send a variant of the situation out into the universe.

Loss then becomes an opportunity for gain. Greater loss then is a "house cleaning" for even greater and better gains.

It is never enough to have just a strong desire to achieve a successful outcome you must also have a strong will to know that it is not only possible but inevitable.

Your imagination is much more important to your destined success than you may have previously understood.

Once the portal to thinking is opened, a inundation of inspiration and new impressions begins to gush out.

Changing Into Something Else

Metamorphosis may seem an unusual word to use in association with personal growth and healing, but the process of a caterpillar changing into a butterfly is one that encompasses what takes place within us as we leave behind patterns of psychological misinformation and move into new areas of development.

The spiritual aspect of changing completely one's life and outlook are too often overlooked by the secular world.

As though there can be an effective cutting away of the soul and spirit from the physical form of our existence on earth.

The work of metamorphosis is expressed through a change in our mode of being.

It is a movement from who we are, to who we can be.

The difference between a human who is going through a Metamorphosis, as compared with a caterpillar turning into a butterfly, is that a human can fully influence the duration of time needed as well as the procedure that induces the spiritual and physical transformation.

The birds that once ate the caterpillar don't even recognize the butterfly that

Now is also flying high in the sky.

The caterpillar has a new identity.

A new beginning.

A fresh start.

Your reaction is under your control.

In any life situation you are always responsible for at least one thing.

You are always responsible for the attitude towards the situation in which you find yourself.

Your attitude is your reaction to what life hands you.

You can have either a more positive or a more negative attitude. Your attitude is completely under your control and can be changed.

Each day you will have the opportunity to learn lessons. You may not like the lessons or think your assignments are pointless and meaningless but each one has a purpose that is special for you and your life.

Imagine waking up EVERY day knowing that you are going to have a day supplied with joy, exhilaration, delight and accomplishment!

Envision going to bed every night realizing fully a REAL peace with yourself and others reliving the marvelous results of your day simply to wake up totally refreshed to do it all over again!

IF you want THIS to be your regular existence, it can be.

Most people on this planet are not happy. Which is very sad actually.

Millions of people are just so depressed and most of them seem to swim aimlessly in a sea of negativity day in and day out, month after month, year after year.

So many depressed people simply accuse the world of all that is wrong in their lives, not knowing what to do to feel happiness.

And even more astonishing is that there are even those who seem to thrive on being unhappy! You know, the person who loves to let you know about their ever-building list of problems.

Happiness is a choice YOU have to make.

It is a claim that only you can establish.

In life little things really do count. You might be tempted to dismiss them, but they are the seeds of daily joy that grow up into the garden of your life.

Not all little things are good, but yours can be … if you are willing to make a little extra effort.

Finding happiness is like finding yourself.

You don't find happiness, you make happiness.

You choose happiness.

Positive and negative thinking are both contagious. All of us affect in one way or another the people we meet. This happens instinctively and on a subconscious level through thoughts and feelings transference and through body language.

The deep truths of existence and life are not found simply by compiling heaps of facts from old books.

Instead, they are discovered in the heart by cultivating an interior stillness from which vantage one can contemplate your mind, your life and the world.

ATTITUDE, BELIEFS, THOUGHTS AND CHOICES

LIFE IS WHAT YOU MAKE IT

Yes, you have heard this before, but have you really thought about what it means? This is a big one.

Your attitude will really make your life.

Nothing, no nothing at all really happens to you. Because you are the one who makes your life.

Yes, in everything.

How?

Your attitudes, beliefs, thoughts and choices.

Take stock today right now and think about it, if you were to die tomorrow would you still do the things today you thought were important?

Think about it and make your life better with every passing moment by having a different attitude, thinking about life in a different way and making different choices. Redefine your definition of success to one which includes compassion for all beings, spending time with family and being more relaxed about life.

You work so hard --- but are you also making time to enjoy life?

Are you getting burned out?

Have you reached a breaking point?

Are you chasing success or doing so much for others that you don't have time to enjoy your life?

Slow Down. Life is simply too short.

Take steps to stop and enjoy the things and people around you. You can learn to minimize the pain of change by strengthening your SPIRIT and to increase your faith to endure and adapt to life changes.

GET USED TO CHANGE ---CHANGE HAPPENS CONSTANTLY

Change really is a true constant in the universe and what is here today may or may not be here tomorrow.

So stay open to learning.

It's always wonderful to learn new things.

Sometimes it's not so great if they are things you should have known.

As I get older, I realize I've learned some lessons that I wish I had picked up sooner in life. Most of these lessons were learned through experience, many through the various mentors I've had along the way and a few from various books I've read. I have also learned from travel to different geographic locations.

What I've noticed is that negative people all over the world tend to adopt a defeatist outlook while positive people focus on creating opportunities.

This is apparently true every place in the world.

It's also interesting to note that you don't see a lot of successful people with a negative outlook on life.

People who are focused on the negative almost always find a way out of pursuing interesting opportunities for success in their lives.

If you spend a little time putting a positive spin on all your actions, it will become second nature, and you'll soon find that you're more productive.

How, you ask?

By being in the NOW.

Every moment you spend worrying about the past is a moment you could spend learning from your mistakes so that you can become a better and more successful person.

Live on purpose and NOT by accident!!

Joyfully alive and being really will connect you with your purpose and the people who you are supposed to meet.

ATTITUDE

It takes strength and courage to transcend the betrayal of those you consider friends, who yet failed to act like it.

The difference between the victim and the survivor is that the survivor gains wisdom and expertise—not just to help themselves, but to help those you will meet later on in life who may not be able to speak for themselves.

I have learned many life lessons, and I am still learning every day. Some of these lessons were from just living one day at a time.

Twenty four hour increments can be life changing.

TIME ONLY MATTERS IF YOU DON'T USE IT WELL

To late I learned to manage my time.

Often I was overwhelmed with social commitments and activities and never felt like I had time for myself.

People were always stressed out and racing to get things done at the last minute. I learned that staying up late to catch up on things, and the lack of sleep made me even less productive in the days that followed.

I found that managing your down time is just as important as managing your up time.

Don't waste your breath proclaiming what's really important to you.

How you spend your time says it all.

Take time to figure out what the "winning" outcome is for you, then work toward it, and do not complain.

Complaining about the heat does not reduce the temperature.

Instead of whining about all that's negative, let's start looking for the positive. The selfishness that seems to engulf everyone at times colors your view so you can't see all the good things around you.

Being grateful is a great attitude.

We have so much to be thankful for, no matter how you make a living or who you think you work for.

The reality is you only work for one person, yourself.

Likewise, you only have two products to sell, your time or your knowledge.

Knowledge alone is not power!

The implementation of knowledge is power.

How knowledge is organized, packaged, presented, shared, or withheld, and even received by others is usually what makes knowledge so powerful.

Knowledge of yourself is priceless.

I do believe that consciousness can exert a form of control over reality.

There is a theory that the original Reality is created in such a way that everything is in divine order with everything else. But in a simulated reality, technology has made synthetic events and materials. This theory states that synthetic objects are really just thoughts created by and through an experiment. They don't exist in the original Reality, so it might be hard to understand that they're only thought forms.

This then complements what the Ancient text teach that thoughts all come from the mental plane, or from a higher-level dimension, and slowly filter down through the dimensions until they get here in the third dimension.

The astral plane is another dimension of this life.

It's one of two planes of the fifth dimension; the other is higher and is called the mental plane. The astral world is of the emotions; the mental is of mind.

The fourth dimension is time. The fifth dimension is beyond time and it is eternity.

This theory is very helpful in trying to understand why we are all so very different.

If you think something is "possible", you can calibrate your consciousness to the frequency of reality in which it naturally occurs. When a new concept enters your thinking, it is difficult to understand or seems impossible. Yet if you do not change your consciousness, you will consider all of these things impossible.

So how do you go about changing your consciousness?

Ancient religions, philosophies and reasoned viewpoints about our life on earth all say the same thing. The universe is made up of energy.

Thoughts are energy.

A person will never rise above the internal images he believes about himself.

☐If you are in harmony with yourself, you can meet a lion without fear.

Start to set your fears aside and open your mind to get to where you want to be. This is usually not anything like what you have previously imagined !

How you get there is immaterial—the Universe will arrange everything perfectly. It will unfold and things will line up perfectly. But you must see it as NOW.

Living one day at a time.

The multiverse is an infinite layer of possible universes. When you fully begin to accept the idea that change, real change and lasting peace is possible, even if it is somewhere else; you have only begun to see the possibilities.

Real metamorphosis occurs by the only way possible- living one day at a time.

www.ingramcontent.com/pod-product-compliance
Lightning Source LLC
Chambersburg PA
CBHW081903170526
45167CB00007B/3125